Jingle in Your Pocket

Dr. Angela K. Love

Disclaimer and Image Credit Notification

Monetary Mindfulness is not affiliated with the United States Mint, the Department of the Treasury, or the United States Government in any way. The United States Mint, the Department of the Treasury, and the United States Government do not sponsor, endorse, approve, or is associated with Monetary Mindfulness.

All United States coin images used in this book are from the United States Mint.

Angela K. Love
P.O. Box 934
Black Hawk, SD 57718

Previously published December 2016 by P-A Financial Services. Previous published ISBN: 978-0-9978407-4-2

Book cover, background, and Penny character designed by Lisa Kerner.

ISBN-13: 978-1-7342305-0-5

Hi! I am Penny!

One day, I heard a jingle in my Grandpa's pocket. I asked him, "Grandpa, what is making that jingle in your pocket?"

He responded, "Coins!" He taught me about all the coins in his pocket.

My Grandpa had pennies, nickles, dimes, quarters, and one half dollar. The coins were shiny and different sizes.

Grandpa told me, "Penny, the United States Mint makes all the coins in the United States."

I learned there are four United States Mint locations: Philadelphia, Denver, San Francisco, and West Point. A *minted* coin is a coin that has been manufactured.

Let's learn about each of the coins!

Every coin has a side called the *front* or *heads* and a side called the *back* or *tails*. The front side of coins have the head or face of one of the United States Presidents.

The back side of coins do not have a picture of a person.

The penny is copper in color and material. The penny was first minted in 1793.

The penny is a one-cent coin. It can be written as $0.01 or 1¢.

This is the front of the penny.

The front side of the penny has a picture of United States President Abraham Lincoln.

This is the back of the penny. The 13 vertical stripes represent the first 13 colonies of the United States.

The nickel is shiny and silver in color!

It is bigger than a penny!
The nickel was first minted in 1859.

The nickel is written as $0.05 or 5¢.

The front of the nickel has a picture of United States President Thomas Jefferson.

Some nickels have an Indian Head or an American Buffalo on the front.

The back of the nickel has a picture of President Thomas Jefferson's house which he named *Monticello*.

The dime is shiny and silver in color like the nickel! It was first minted and used in 1946.

The dime is written as $0.10 or 10¢.

It is the smallest in size of all the coins!

The front of the dime has a picture of United States President Franklin Roosevelt.

The back of the dime displays a torch representing liberty.

An olive branch is to the left of the torch, indicating peace.

An oak branch is to the right of torch, signifying strength and independence.

The quarter is shiny and silver in color like the nickel and dime! It was first minted and used in 1932.

The quarter is written as $0.25 or 25¢.

It is larger in size than the penny, nickel, and dime!

The front of the quarter has a picture of United States President George Washington.

The four U.S. Mint locations have created many different designs on the back of quarters. The design on the back of the quarter on the right is part of the National Park Series and is a picture of two bats.

The half dollar is shiny and silver in color like the nickel, dime, and quarter! It was first minted and used in 1792.

The half dollar is written as $0.50 or 50¢.

It is the largest in size of all the coins!

The front of the half dollar has a picture of United States President John F. Kennedy.

The back of the half dollar is the Presidental Seal. Fifty stars surround an eagle with a shield on it's breast holding 13 arrows and an olive branch.

We have learned about the penny, nickel, dime, quarter, and half dollar.

Let's have some fun!

Point to the coins that are not the penny.

Remember, the penny is a different color than the other coins!

Great job!

None of the coins below are the penny.

Can you point to the coins
that are not the nickel?

The nickel is almost the same size as the penny!

Way to go!

These are the coins that are not the nickel.

Can you find the coins that are not the quarter?

Point to all the coins that are not the quarter!

You are a hard worker!

All of these coins are not the quarter.

Which coins are not the half dollar?

Remember, the half dollar is the biggest coin!

Excellent work!

These are the coins that are not the half dollar.

Now, we need to pick out all the coins that are not the dime. Which coins are not the dime?

Did you remember that the dime is the smallest coin?

You did great!

The dime is not among these coins.

My Grandpa showed me how to combine coins so that they equal each other in value. I will show you how it works.

Let's learn some coin combinations!

Five pennies equal one nickel.

Look at the combination of coins below. One dime is equal to five pennies and one nickel.

It is fun to learn about coins!

Did you know 10 pennies
equal one dime?

=

That is a lot of pennies!

I did not know that one quarter equals two dimes and one nickel until my Grandpa showed me.

Take a look!

You are doing a great job learning about coins!

One quarter can also equal
five pennies and two dimes.

I did not know there were so many
combinations of coins. Did you?

Did you know that one quarter can also equal four nickels and five pennies?

Whoa! There are a lot of combinations of coins for one quarter!

Here is one more combination for a quarter. One quarter equals 25 pennies! Can you count all 25 pennies?

=

Can you help me identify coins? Which coin is the half-dollar?

Finding coins is fun! Thank you for your help!

How about finding two pennies?
Can you help me find two
pennies by pointing to them?

I like counting pennies! You are a big helper!

Let’s find three nickels! Can you point to three nickels?

I found the three nickels!
Did you find them?

I need your help finding four dimes.
Can you help me?

You are good at this!
Thank you for your help!

I need your help once more! Can you find five quarters?

We did it! Counting coins was so much fun!

Let's try a new game! Can you place the correct coin in the correct circle?

Half Dollar

Quarter

Penny

Hurray! You did it!

Dime

Nickel

How about now? Can you place the correct coin in the circle where it belongs?

$0.25

$0.05

$0.01

$0.10

$0.50

Great work!

Here is another you can try! Place the correct coin in the circle where it should go!

10¢

50¢

5¢

You did a terrific job!
Thank you for all of your help!

1¢

25¢

I had fun showing you what my Grandpa taught me!

I hope you had fun, too!

Made in the USA
Monee, IL
05 November 2026

67379441R00024